A JOURNEY TO THE OUTER LIMITS OF THE MIND

THE

DÉJÀ VU

EXPERIMENT

J . G . RENATO

Veritas Shield™

Paperback ISBN: 978-0-9897186-1-5
Hardback IBSN: 978-0-9897186-0-8

Library of Congress Control Number: 2013945329

PRINTED IN THE UNITED STATES OF AMERICA

The Deja Vu Experiment

Reviewed by Patricia Reding for *Readers' Favorite*

What might John Galt have thought if he had left his followers in the valley of Colorado? Would he have continued to follow the idea that only the material mattered? Or, might he have undergone a philosophical revelation? In The Déjà Vu Experiment, J.G. Renato expounds upon this idea. The Déjà Vu Experiment is not a "story." It is, rather, a discussion of the little anomalies in life that may lead one to look at the world in a new way. These anomalies, these "gaps" were discovered and then examined by John Galt when he met Diana, an Iowa farm girl who encouraged his curiosity.

Renato suggests that people are so intent on their own internal realities that they fail to notice the greater world. He challenges them to "look through the veil" of their historical understanding. It is the strange little events that will wake people from their typical hypnotic approach to life, events often brought forth through art, theology, and science. Living our "mortal dreams," we miss out on an appreciation of the eternity in which we live, an eternity without past, present, or future.

I especially appreciated the quotes Renato shared from Marianne Williamson, Albert Einstein, Jack Kerouac, the Chinese philosopher, Zhuangzi, Tolstoy, and more. Some of his ideas encouraged my thoughtful consideration and challenged my understanding. For example, Renato suggests that if we operated just as spirit, if we knew of our immortality, "it would be tough to get a rousing game of life going." He also asks, which is really in charge: you or your body? Offering unique ways to look at light, quantum physics, string theory, the universe existing as a single unified melody, the power of imagination, free will, the language of mathematics, death, and more, Renato successfully challenged me to consider not just "Who am I?" but "What am I?"

CHAPTER ONE

———————— ∞∞∞ ————————

So you think you can tell…
A smile from a veil
…Do you think you can tell?

—Pink Floyd, "Wish You Were Here"

I f you're like me, you've noticed these little red flags in the physical universe. The ones that hint all is not what it seems. Maybe you haven't acted on them—I didn't for a long time. I was too busy stopping things. But these little flags, these little anomalies, have caught your attention. Made you wonder. Made you think about God.

They might have made you uneasy for a millisecond, but if you're like most people, you find it easy to go back to accepted reality. To go back to living what might actually be "the dream." Looking at life through what might actually be a "veil."

Here's what I mean. Ever been driving along an interstate and suddenly you realize you have no real recollection of the last thirty miles? It's a little unnerving. You have been so intent on your own internal reality that you haven't noticed anything about the reality around you: the green exit signs, the towering cumulus clouds to your right near the horizon, the Bruce Springsteen song on the car's radio. You weren't aware of any of that.

It's spooky.

But in two seconds you dismiss that peculiarity and you're "driving" again.

What's spookier yet is that a good portion of people's lives may be lived like those thirty miles on automatic pilot.

But there are these little red flags you occasionally notice about the apparent "reality" of the physical universe. Strange little events that wake you up momentarily from those hypnotic thirty miles on the interstate. Little moments of apparent discontinuity in space or time. Non-sequiturs in the expected ebb and flow of the physical universe around us. Unnerving moments when we "awake" for just a second, but that we immediately dismiss.

This book is about *not* dismissing them.

The first time one of these moments happened to

me was one afternoon when I was about ten. I had gone down to the garage my dad ran in the small town in Ohio where I grew up. I was a lot like my father then. He was a mechanic, a technician, and he measured himself only by what he created, not by the outward image he projected to others.

He and I had a soda from the pop machine, and then I went around to the back of the garage. I walked across the lot, littered with a dozen automobile carcasses in various states of disrepair, and I sat down on rough grass on the top of a bank next to the railroad tracks. I did some of my best childhood meditating there. Something about that spot brought out my curiosity about the world.

That afternoon I focused on a tall grain elevator about a half mile away, in a field on the other side of the railway tracks. It was about a hundred feet high and it was a flat tan color. The name CASE, painted in huge black letters, circled its flank.

The more I looked at it, the more I felt I could "connect" with it. Then, the more I started to think about its physical presence in its actual location—as if I was no longer viewing it from a distance—the more I could "feel" it. As opposed to seeing it through my body's eyes as an object distant from me, I could feel where it was and where I was. Simultaneously.

As I did this, the physical universe around me became quieter and quieter.

Then, in one moment—which could have lasted no time at all, or forever—I could experience with certainty both the exact location of my body and of the tower, like two poles in one of my dad's car batteries.

All "normal" appreciation of distance and separation vanished for me in that split second.

All "normal" perception of the physical universe as a group of solid, heavy objects ceased.

The world seemed more like an illusion. Light. Airy. Playful. Something imagined. A creation. A thin veil.

It took my breath away.

You should try it sometime.

This experience was the physical equivalent of déjà vu, that simultaneous experience of the past and the present. It's an anomaly in time that "pops" you out of the stream of an otherwise sequential reality.

That afternoon, for me, it happened in space, not time. But it still got my attention.

You've had the déjà vu experience happen to you, I'm sure, without even trying for it.

These little contradictions between the chronically expected and the temporarily perceived reality of space and time around us are what I call "gaps." And just as the Deconstructionists use gaps—the contradictions in the surface of a text, or *lacunae*, if they wish to sound more technical, more academic, more authoritative about it—to begin to explore the truth or ultimate reality behind a text, so too can these gaps in the space and time of the physical universe be used to follow Alice down the rabbit hole, so to speak.

The idea that the world around us is an illusion, a veil, a tabernacle, maya, is hardly a new idea. Religions have been telling us this metaphorically for centuries. Well, perhaps not so metaphorically—but without an actual, personal experience of the divine, and without any hard evidence, it has seemed pretty metaphorical to many. For most of us living in the veil, those thirty unconscious interstate miles don't even register as a metaphor for something larger. After all, here in America we're red-blooded, rugged capitalist individuals. We prefer to see our world as a world without gaps. A world of technicians and producers. An utterly rational world.

We hope.

But Zen koans and New Testament maxims alike lead us toward those little gaps, not away from them. They call for us to take notice of momentary discontinuities in our perception of life as we know it. "What is the sound of one hand clapping?" and "It is easier for a camel to go through the eye of a needle than for a rich man to enter into the kingdom of God?" prompt us to question our conventional, logical, and rational view of physical reality and, if taken seriously—with religious sincerity, even—serve up these little gaps to us on a metaphysical platter to tempt our intrinsic spiritual appetites.

Artists through the ages have also been intrigued by these discordances in the universe around us. René Magritte's *trompe l'oeil* paintings and M. C. Escher's sketches of impossible geometries momentarily pop us out of our perception of reality and tease us into thinking about physical reality from a difference perspective.

And that's the crux of this. Are these gaps really just tiny anomalies in the expected flow of time and space around us, easily and appropriately dismissed because they are meaningless illusions? Or are they actually the keyholes through which we can glimpse a greater reality, the eyes of needles through which we can pass, metaphorically or actually, to experience what might really be heaven?

Are they the first steps on a path to what might really be the ultimate, *ur*-religious experience? The ultimate knowledge?

I had been living my life—very successfully I thought, very nobly, even, and very rationally—as an engineer and designer. I was someone on a mission "to stop the motor of the world," as I so less-than-humbly put it. But then the arrest after my television address happened, and it was only Diana, an Iowa farm girl, a beautiful widow, who defused my need to be so right about everything and encouraged me, instead, to return to being curious, as I once had been as a child.

When I had regained my curiosity about life, she taught me—even more important—how to look at life without blinders, so I could actually see it.

I had met her when I was working the rails for Taggart. She was a waitress. She never did more than serve me coffee and I never did more than drink it, but we could see in each other's eyes the seeds of a great closeness. I probably was a closet romantic, even then, although I would have denied it left, right, and center. When I separated from the others after they left the valley in Colorado, I went to visit her. Somehow I knew she could help me sort out what had by then become my confused life.

When I arrived back in her little Iowa train stop of a town, I found out that her husband had been killed a year or so before when the tines of a harvester had scraped the life from him. They had no children.

She had the keenest intelligence of anyone I had ever known…and she knew exactly when to display it and when not to. We had married almost immediately after I'd arrived, and we stayed together more than twenty years. She died in her late sixties. She was my second love. By that time, thanks to her, I had found myself. Even though she was gone, I still felt whole. Diana woke me up.

She got me listening to Pink Floyd, not just Bartók and Halley's Fifth Concerto. She got me looking at David Hockney, not just engineering blueprints. Ever since I had left my father's garage at twelve, I had been driven. But I was not driving. I was on automatic pilot, just like those thirty miles on the interstate that you, too, have done without seeing a thing. I was not awake.

I learned from her that I had to stop being so serious, so intense. I mean, the game had gotten pretty serious for me. She lightened me up. She got me to confess my secret experience out behind my father's garage—I thought I must be crazy, so I had never told anyone about that. But she got me to talk about it and to remember what it was like to feel "above it all." I was childlike and curious again. I, my true self, could see.

She taught me to notice the gaps.

Over the ages, theology and art have provided the bulk of these invitations—these gaps—for us to follow into a greater reality. Today science, too, has begun to thoroughly explore these gaps. It uses them to posit an underlying "theory of everything" that may seem to border on the mystical, but is in fact based on observable physical phenomena.

I'm talking string theory and superstring theory, a photon that's simultaneously both a wave and a particle, a resolution for the dichotomous twentieth-century theories of general relativity and quantum mechanics, a solution to the arrow of time quandary, the immeasurable infinity/nothingness of the singularity at the center of a black hole.

I'm talking the revelation that the physical universe is a diaphanous, vibrating veil where the terms "past," "present," and "future" no longer have any meaning.

I'm talking about eternity.

Yes. I'm talking about you and me. Immortals. But living our identities in a mortal dream. I'm talking about *what* we are, not *who* we are.

Why did Alice go down that rabbit hole, anyway? What gap was she hot on the trail of?

The Greeks admonished us to know ourselves, encouraged us to find out what we wanted. Alice was on a journey of self-exploration, too. That's a journey you and I could begin if we could ever stop ourselves from shying away from the frightening potential that lies within the gaps we detect.

Diana helped me shed my fear and let me know that fear comes from being immortal. Immortality, after all, does imply a formidable responsibility.

Oops. Got serious there for a second. Forgot that it is actually a freedom, a joy, to wake up from those thirty miles of automatic pilot on the interstate of life.

There are big questions that we tend to shy away from: What are we? What are we here for? Do we have free will, or are we merely living out the reality created by some other entity—divine *or* diabolical? Is there any hope for us amid all the worldly suffering we see? Why would anybody create suffering to begin with?

All good questions, if we dare peek through the veil

to find the answers. If we can tell a smile from a veil…
if there's still a smile to be had after we see what the veil
is actually made of.

Just a joke. Sorry. My rehabilitated humor is still a
little bit wry.

Let's dare to follow Alice down that rabbit hole.

CHAPTER TWO

*The most exciting attractions are between
two opposites that never meet.*

—Andy Warhol

Everything starts with contrast.

We know "up" because of "down." We know "dark" because of "light." "Cold" because of "hot." We only know one thing by comparing it to another. "Up," "dark," and "cold" are meaningless by themselves.

If, in the beginning, there was only one speck, one particle in the whole universe, you wouldn't have been able to tell if that particle was moving or if it was stationary. If there were two particles and one *appeared* to be moving, you still wouldn't have been able to tell which was moving. If *neither* particle appeared to be moving, you also wouldn't have been able to tell if they were both stationary or if they both were moving at the same rate.

Comparison precedes knowledge. Contrast rules.

The gaps come in between.

My days were defined by contrast from the time I left home at twelve. I learned early that the rational world valued logic. I had to operate within its rules when I studied math, engineering, and physics at Patrick Henry University. The application of force in one area caused an equal and opposite reaction in another, and so on. Then I applied that rational thought to my design for a static electricity engine for Twentieth Century Motors. But my view of the world as a place of contrasts was locked in when I decided to stop the motor of the world, to stop "them," the ones who were the opposite of "us."

Like everyone else—except Diana and a few other insightful souls—I came to understand the world by the opposites I and others observed within it. Language was the tool we used to label these opposites, I knew, but I was also sure they existed without the labels.

Or did they?

What if we looked at these "opposites" not as part of a pair, but as individual items with no other significance other than their own integral qualities? Or could we even do that?

Our minds, which are shaped by language, function only on the principles of opposites—of logic. They label dark and light, up and down in the physical universe, yes, but they also label strong and weak, love and hate. They append qualities to quantities and actions. They can make opinions overlay the facts.

This object is beautiful, that one is ugly? Why? It's pure opinion. It's the placement of a subjective value over an objective fact: both objects are merely objects.

That entire spectrum for defining our existence, from fact to opinion, is measured in opposites.

We lay a veil of qualities over quantities that are already there around us. Plain, simple, individual quantities.

The gaps come in between.

Our minds strive to understand the universe through a framework of contrasts, not by simply perceiving it—as I was briefly able to observe that tower and my own body behind my dad's garage.

Because of the way our mortal minds are taught to function, Zen koans work. Take "What is the sound of one hand clapping?" When we try to "think" about this, we feed energy into the mind. That energy reverberates between the two opposite "idea" terminals the mind has created in order to "think" about a contradictory concept. We are taught that sound must come from two objects colliding, but one hand is only one object. The statement must be a trick. The interesting part of the question occurs between what we have observed and what we believe we know—it happens in the "gap" between them. Feeding enough energy into the mind to think about something that seems to be an inherent contradiction eventually short-circuits it. Diana taught me the benefit of doing just that, though…with a little help from Alan Watts. A little electrical meltdown occurs, just for an instant—like creating a battery with two positive or two negative poles—and we experience the gap in the veil.

It's the same mechanism of temporal or spatial contradiction used in René Magritte's *Les Valeurs Personnelles*, where vase of a blue sky painted with white clouds sitting next to an armoire looking out on a blue sky with white clouds. It "pops" us out as viewers. It provides us a little shimmer of aesthetic titillation as we enter the gap generated by our physical senses' inability to resolve the visual contradictions. It's a *trompe l'oeil*. It tricks the eye…and the mind, and therefore us, too. We "pop" out. Of our bodies and our minds.

That's also why the cascade of silence that follows the tension of violins in Samuel Barber's *Adagio for Strings* suddenly removes us from the climactic build-up of energy flowing between those strings and "pops" us out for a second into a gap far more silent than the simple absence of noise that exists before the final violin resolution begins. The polarity between noise and silence itself gets our attention.

That's why the forty-two-second-long final note of The Beatles' "A Day in the Life" "pops" us out and causes us to vibrate in the gap between our expectation of that final note ending and the reality of it not ending.

That's also why we "pop out" when viewing M. C. Escher's infinite staircase in *Relativity*.

These opposite poles permeate our inner and outer worlds.

For any energy to flow—for there to be any "life," in other words—there must be two poles for energy to flow between. One positive, one negative. One transmitting, one receiving. No matter the significance that the human mind places on each pole, the poles themselves must exist for energy to flow.

From the yin and yang of the East to the war between God and Satan in the West, contrast between opposite forces always precedes the game of life. "All the world's a stage," and that stage is made of contrasts.

Let's face it: You're stuck in one of those contrasts right now, one of those opposing polarities, while you're sitting there reading these lines.

You've got a body, male or female. The core physical contrast between all humans.

And the one most necessary for the game of life to continue on this planet—through procreation. Whether understood as a yin/yang of sperm and egg or as the placement of Adam and Eve by God in the Garden of Eden, we need both complementary forces to complete our biological evolution, for companionship, and for what has become an endless begetting of offspring. Either way, contrast is key.

That body you're sporting is one half of what's necessary for the future of human life on earth. It exhibits the same male/female polarity that drives the procreation of all the animals on our planet—all the bears, bees, and bass. It is what's necessary for "the force that through the green fuse drives the flowers," thank you very much, Dylan Thomas. A force must have two terminals to flow between, otherwise there's no flow. There's no life.

The energy that flows from quark to quark as a weak or strong force, that flows from planet to planet as gravity, or from sun to sun as electromagnetic waves all flows from one pole to another.

Even if we don't yet really understand how.

Physics in the last hundred years has given us far more of an insider's look at our universe than Galileo or Newton could in their time, but theories of relativity and quantum physics still run right up against each other. Yes, more contrasts that arise from trying to explain how the physical universe works. Einstein discovered gaps in Newtonian physics and now we're discovering gaps in Einstein's physics. While theories of operation for the "small" forces of electromagnetism and the strong and weak nuclear forces all merge into a satisfying harmony with each other, the theories about the fourth, "large" force of gravity remain diametrically opposed to the other three.

Is this opposition because of the way the physical universe operates or because of the way the human mind observes it?

In either case there remains a contrast, and between

those poles is the gap. Twentieth-century theoretical physicists are straining to arrive at a "theory of everything" that will unite these discrepancies, these polarities, and, perhaps through the postulated extra dimensions of string theory and superstring theory, fill in that gap.

Or not.

Meanwhile, you're sitting there in a blob of something that's not much more than a piece of animated protoplasm whose closest kin are bears, bumblebees, and broccoli.

Exactly who are you? That's what the caterpillar asks Alice. And how do you get to "know yourself" in this endless panorama?

Maybe the caterpillar should have asked Alice *what* she was instead.

What is it that allows you to "pop" into these mini out-of-body experiences?

This is the question that Diana, somehow, knew the answer to.

Look. Stage left. Enter the human soul. A nice round of applause, please.

And what is it that you "pop out" of? Enter the veil, stage right.

The soul and the veil. The Big Contrast. Spirit versus flesh. Immortality versus mortality. True knowledge versus ignorance. Nothing versus something.

The *final* contrast?

Maybe. Maybe not. Here's the Big Question: Are you a "thing," or are you a "no thing?"

The first step to take in answering that question is to take a look at what we are doing living in a veil in the first place. Why the mortality? Why our infusion in these fleshy tabernacles? Why all this worldly suffering? Why do we have to hang around in these "vessels" until disease, old age, or a speeding Toyota ends their usefulness?

A body imposes more restrictions on a spirit than a fifty-mile-per-hour governor on a U-Haul van. Believe me, I know. I know motors.

Anyway, here's the point: If we operated only as a spirit, if we knew we were immortal, if we were all-knowing, if we had no restrictions, it would be tough to get a rousing game of life going. We would always know the outcome, and there could therefore be no game. Worse yet, there would therefore be no fun. Any game would be over before it started. Of course, if we're going to have fun…we also have to have its opposite. Contrasts rule.

In short, mortality gives us something to do.

To be "mortal," though, to seem to die, we have to don the "governor" of the veil so we can pretend we don't know we're immortal, all-knowing, without restriction. As Marianne Williamson writes, "Out deepest fear is not that we are inadequate. Our deepest fear is that we are powerful beyond measure. It is our light, not our darkness, that most frightens us."

Our responsibility.

After we reduce ourselves to mere mortals so we can have a game, we may no longer be serene and nirvanic beings, but at least we'll have something to do.

Alice goes down the rabbit hole so she can have a game. So she won't be bored.

For Alice and for her readers, the game is immensely fun. For me, the game had become much less than fun, but I only recognized it that night we all flew out of New York City on our way back to Colorado.

As our plane headed west and I looked down and saw the blackness descend over the great city, I could no longer share the satisfaction my fellow passengers felt. Yes, we had done it. We had stopped the motor of the world, but what was it we had really done? What was it I had done? I had become a destroyer. The opposite of a creator. I had mistakenly, unconsciously defined myself as the opposite of "them." I had thought they were destroying me, and so I became a destroyer myself.

It was at that moment that I realized I had also destroyed my own integrity.

I had become *one of them.*

Today I know that it is better to be a pawn in my own game than to be a king in theirs.

But then, I and all those who I had recruited away from the outer world had so exalted the importance of the mind that we came to believe that no sane person had ever believed in anything irrational or ever tolerated anything not logical, anything that could not be analyzed by the mind's reliance on contrasts. Our spectrum had the notion that history had made the mind evil on one end and the mind as a purely logical organ on the other end.

We had embedded ourselves on one end of that spectrum. To us, the mind was God, the mind was not evil. By doing this, we had blinded ourselves from ever being able to notice a gap somewhere in between.

I had latched on to one extreme rather than look at the whole problem of the mind—as either God or Satan—as the actual problem. My own refusal to believe that I had been born with any original sin just by having a mind, or to tolerate the idea that the use of my own mind to attain happiness was immoral led me, mistakenly, to the idea that the mind was all.

Like Narcissus, from Greek mythology, who was born with a body so perfect that he began to worship it, I began to worship my own mind.

Worse yet, I had persuaded others to do the same. On that flight back to Colorado from New York, I knew all life had become darker for me than the streets of New York City below.

CHAPTER THREE

There is a crack in everything.
That's how the light gets in.

—Leonard Cohen

Perhaps my darkness was why Diana began sending me to the library to read everything I could about light when I first moved in with her.

Light. She said it was the quintessential example of contrast. She said it was a portal into the quintessential gap. She was right.

It behaves as both a particle and a wave, per modern physics. It's a polar contradiction in itself that points to the divine. Light has been used to depict the ethereal—that which is behind the gap—in sculpture and painting in the East and the West throughout the history of the planet.

Light is unique.

Not only does it give rise to life in corporeal organisms, it regulates their daily cycles as well through its impact on circadian rhythms. It can pass through the smallest of cracks in solid objects, such as a curtain near a window to light up a room on the other side of the wall. So too are ghosts or spirits said to pass through and across solid barriers in space, as when they reassure distant relatives that their death was only the start of a greater journey and that there's nothing for their loved one to fret about. Sometimes they even appear to reassure in the guise of a filmy, light-like aura.

No wonder light has historically been seen as a residue from some higher dimension and has been used to imply the realm of spirits, angels, and even God. Light, better than anything else, shows "the ghost in the machine."

Take the halo or aura, for example. It's associated with the spirit behind the veil, the divine within the tabernacle of the body. From ancient statues of Buddha to portraits of multilimbed Tibetan deities to depictions of the Egyptian god Re, the halo has been represented in art as the light of the spirit. It generally sits behind the head of "evolved" figures. Sometimes that light even takes up the entire space behind the body on a canvas or wall. It stands for a big, evolved being—the dreamer behind the dream. Light is associated with the divine nature of our essence, our native presence.

In the West, the light of the halo has adorned religious figures from Christ to Saint Francis since the fourth century and has been painted as everything from a cumbersome square behind the head of Pope Paschal I in a mosaic at the Basilica of Santa Prassede in Rome to the delicate, transparent circle behind the head of the Madonna and Christ child in Leonardo da Vinci's *Benois Madonna*.

But wait. In the West, the Catholics—bless their Inquisition-minded souls—have gotten it wrong. They believe that people *have* souls, and, literalists that they must have been, they created paintings that make it seem like a person's soul is somehow attached to the body like a Prada purse or set of noise cancelling headphones.

Oops.

Let's make a 180-degree change in the idea of who's really in charge, you or your body. As C. S. Lewis said, "You don't have a soul. You are a soul. You have a body."

Here's the proof. You have to actually do this little drill so you can experience it, experience, in fact, the veil in all its "glory," for yourself. Otherwise all that is in this book remains merely a collection of cute, abstract ideas, not a reality, and you will remain a part of the collective dream, not start to become the individual you really are.

Try this. It shows that *you're* the halo, *you're* the light. Light isn't the portal to the divine, *you're* the light at the end of the tunnel, looking back down into and across the veil.

An acquaintance who dabbled in Eastern philosophy first showed me this when I was double majoring in physics and philosophy at Patrick Henry University. I dismissed it immediately because there was no scientific proof that what I had just experienced was valid. Diana, of course, had already been doing this for years before I met her, and she and I each "did" it day in and day out to keep ourselves, as best as possible, out of the "veil."

Here's what you do.

Look at the wall across the room from you. You can perceive it, right? One of the reasons you can is that you're not the wall. You are separate from it, either in quantity—the distance between you and the wall—or in quality—the nature of you as opposed to the nature of the wall. You can only perceive something if you're not it.

Then close your eyes. Draw your attention to your left ankle. Then focus on your right elbow. Then on that large bone on the top of your shoulders that we call a head. You can perceive all that, right?

Guess what? You're therefore not it. You're not your body.

Okay. Step three. You had something to eat or drink when you first got up this morning, yes? A cup of coffee or glass of juice, a bowl of cereal or yoghurt. Okay, now close your eyes. Return in time to the moment you drank or ate that. You're looking at a memory of the event, a picture of the room you were in and of the coffee cup itself or the bowl of cereal, right? You recorded that memory and it's available for you to look at again, any time. It's part of your mind.

Bingo. You're not your mind either.

You *have* a mind. You *have* a body. They're both different in quantity or quality from *you*.

You're the halo. You're the light. You're the spirit.

The halo. In what dimension is that? The dimension of "you," perhaps.

I had set myself to doing this exercise daily with Diana. It was almost a meditation, but it had more immediate and practical results. I guess I was still a bit of the technician, like my father. But now I knew what, if not yet who, I was.

I was the gap.

That fine circle of light in da Vinci's painting, so distinct from the "material" world around it, is that Renaissance artist's depiction of the gap.

The Impressionists, too, recognized the magical properties of light. They showed them not as halos around holy people, but as an almost intangible presence permeating the world. In Claude Monet's 1891 *Haystacks*, light glows throughout the entire scene. It doesn't just come from a source behind the haystack, it somehow infuses the entire world with a quiet but powerful glow. It's the immortal lighting up the mortal. In Georges Seurat's *Sunday Afternoon on the Island of La Grande Jatte*, his Pointillist technique embeds a tiny source of light into every daub of color that makes up the scene of people, trees, and water and permeates that world, too. Once again, the immortal illuminates the mortal.

Light permeates all.

Perhaps you do, too…from outside the veil, just as the dreamer permeates his dream. Perhaps all of our collective light, our collective "soul-ness," permeates this entire physical world.

Remarkably—or perhaps not, given the possible interconnectedness of that which lies outside the veil, that community of haloed beings, us—while the Impressionists and Post-Impressionists were quietly illuminating their paintings with the incandescent presence of light, Einstein was calculating on his blackboard the mathematical and physical configurations for the same magical "force." His discovery about the properties of light was as spectacular a leap into twentieth-century physics as Monet's and Seurat's capture of light on canvas was a major stylistic leap in art.

Einstein's discovery that the speed of light was not only a constant, but was the fastest constant around— the universal speed limit of particle flow—turned on its head Newton's idea that gravity was an instantaneous force—a force, as it were, without a governor. Newton's instantaneous gravitational travel may not be possible in the physical universe, it turns out. (Only ghosts or haloed beings can do that). Einstein knew that it takes eight minutes for light to travel from the sun to earth, ninety-three million miles, and that gravity, therefore, could not exceed light's speed.

From that came Einstein's theory of general relativity and, presto, space and time were suddenly redefined.

Space-time now equaled the components of the four measurable dimensions, not Newton's three spatial ones, and these four themselves composed a fabric that warped and curved because of the force of gravity. The warping and the rippling of the fabric itself traveled at the speed of light.

Pretty cool. So where's the gap here?

It almost seemed to disappear until the 1920s, when those pesky quantum physicists arrived and announced the discovery of a new nano-dimensional world smaller than protons, electrons, and neutrons, a world filled with quarks, charms, and other minute and improbably named particles.

Enter, stage left, once again, the gap.

Turns out that this newly discovered, teeny sub-atomic world and the theories that explained it meshed nicely with the theory of electromagnetism, but were completely at odds with the rules for gravity and its theories of operation. There was a gap between how the universe seemed to work on the macroscopic level of gravity and on the microscopic level of subatomic particles.

In fact, it was a complete contradiction, a gap large enough to drive a black hole through. The equations and theories that govern the physics of the "large" fell apart completely and made no sense whatsoever when applied to the physics of the "small."

Thanks to Einstein, the calculation of the speed of light—something everyone could agree was constant—opened up a search for a unifying theory of how the universe operates that is still in full cosmic swing today. If that "theory of everything" can be found to explain the gap between the large and the small, there will no longer be a gap between anything, contemporary physicists think.

Maybe. Maybe not.

There's still the halo for them to reckon with.

CHAPTER FOUR

For the listener, who listens in the snow,
and, nothing himself, beholds
nothing that is not there and the nothing that is.

—Wallace Stevens, "The Snow Man"

Maybe, after I had experienced myself as "nothing," as a soul hovering above my father's garage and the grain tower across the field, I had been frightened into always needing to be something. Maybe as exhilarating as that moment was, it had also terrified me. Being "above it all" meant not being a part of anything. Suddenly I became driven to be something, and, being driven, I was out of control and on an automatic-pilot kind of journey to become "something," some thing.

The mind, as a thing, was conveniently at hand.

Later, after I had decided to stop the motor of the

world, all sorts of legends began to form about me. The story about me being an explorer who found the fountain of youth on a mountain top. The story about my being a billionaire on my yacht and spotting the towers of the lost city of Atlantis. Others passed these stories on to me.

I started to become caught up in my own fame. I started to take pride in the apocryphal stories I had heard about me, no matter how far from the truth they were. I started to accept the lavish praise I was receiving about the static electricity engine I'd designed, but I was blind to the fact that basking, even quietly, in that praise was the passive equivalent of boasting. My mind, that glorious tool I had become, really was something, I decided.

I lost what little compassion I had had as a youth for my fellow man.

I began to promote myself, covertly, to those I wanted to join me as destroyers by taking advantage of the fame that was building up around me. I enlisted the best rational minds I knew to help me destroy in the name of creating. Subconsciously, though, I was beginning to doubt my own ability to create any longer. I began to feel like one of those one book authors, like that gal who wrote that book about the rugged individualist architect Roark, and I hid away the other creations I was working on, afraid I might not be able to pull them off.

All those inventions that were destroyed in the back room of my apartment in New York before we exited the city were the machines I had talked about and had added to the widening circle of my fame, but in fact none of them really worked. I destroyed them so that no one could ever find out my little secret.

By then I had become terrified of not being "something."

Only Diana knew this.

I loved the warmth of her upper arms. I'd come up behind her and place my palms flat against her arms and let my chin rest on the thick braid she'd sometimes let fall from her head. Somehow I had known she was safe for me, known she was the one I needed to go to after I realized in the plane over New York what I had become. Somehow I knew she would listen, knew she would help.

She was pretty clever. She got me looking at contemporary physics in ways I never had looked at it before. She knew I puzzled over the contradictions between current theories, but she got me to look at them so that I would appreciate the gaps, not so that I would be driven to keep on solving the problems of discord in them, the problems which had originally called attention to the gaps.

She pointed out to me, for instance, that until recently the quartet of prime forces in the universe—electromagnetism, the strong and weak nuclear forces, and gravity—had not come into theoretical harmony with each other. No matter what vantage point they were examined from, the underlying melodies of each remained fundamentally out of key.

Enter string theory, but not stage right or stage left. It had bubbled up from the minds of contemporary physicists like the ones I had studied at Patrick Henry. It postulates more dimensions—beyond those of left, right, up, down, or even of tick-tock. Suddenly there was the magical appearance of the first subatomic-, subquark-size stepping stones of a path that might lead to a theory of everything.

String theory, in its attempt to reconcile general relativity with quantum physics, asked, "What if the basic building blocks of the universe are not matter, not particles at all, but are vibrations? And what if those vibrations underlie the known physical universe's four dimensions—height, width, length, and time—in ever tinier layers of up to seven other dimensions?"

If the theory of vibrations proved to be true, the universe could suddenly be understood as one unified melody, and every one of those tiny strings would be vibrating in the same key. There would be harmony. There would be an underlying theory of absolutely everything. As Diana fed me more and more of these theories, like any rational engineer, I became more and more excited about finding a solution.

These strings and superstrings were described as minute, one-dimensional squiggles that, like guitar strings, could be stretched and vibrated to produce a unique sound. Physicist Brian Greene noted that if our entire solar system were the size of one atom, an individual string would only be the size of one tree.

We are talking tiny, tiny little oscillating poofs of things that float in space-time and, by the combined symphony of their vibrations, produce the known physical universe and the four forces in it. This was exciting to me.

But Diana was pretty tricky. She knew what she was doing with me.

While I was discovering that string theory and

its expanded offspring, superstring theory, were both mathematically consistent and offered a harmonious theoretical underpinning to both gravity and the other three forces, I found that they were—a slight drawback here—not provable currently through any hard evidence.

Nonetheless, their multidimensional view was as far reaching as any of man's most imaginative dreams. Despite the fact that, to some, it was as unbelievable today as Copernicus's idea that the earth might not be flat or the center of the universe had been in the 1500s, it might not be so impossible.

Diana then continued to guide my research into even more esoteric avenues.

Enter the idea of an increasingly ethereal hierarchy of seven additional dimensions, each a finer vibration of the one below it until the one on "top" begins to be merely an aura—dare I say, a halo—of infinite possibility. Not so far-fetched an idea when we consider the recent discovery that humans who have been physically paralyzed are now able to control a cursor on a computer screen with their brain waves alone. Astrophysicist, philosopher, and chief superstring theory advocate Stephen Hawking is perhaps the best-known user of this technology, but scientists have also been able to teach other primates to use their thoughts to move mechanical arms just as if they were their own.

Could it not be, then, that an energy exists in those other dimensions that is different than the conventional physical universe's energy? Energy as we know it composes the surface of the veil, but yet it can impinge upon the physical universe, as we see from Hawking's and the primates' control of matter through thought? What dimension, for instance, does a dream occur in?

How unbelievable then, really, are the extra dimensions postulated by string theory that underlie the physical universe and yet have an effect on us?

No more unbelievable than a halo, I began to see.

Not to mention that they unify the operations of the four known forces. A scientist's dream. As well as the dream of a haloed being. I and Diana had come into unison.

Perhaps the eleventh dimension will ultimately turn out to be human consciousness itself. The not-easily-measurable field from which all human ideas spring, from which all dreams are created, a limitless reservoir from which all human goals are formed.

A workable theory of everything?

That means there'd be *nothing* left.

Literally, no thing. In other words, me.

I could finally stop worrying about being something.

Just as it took Copernicus's mathematical equations to overturn the belief that we were at the center of the universe—because of what our eyes told us—so too it took superstring theory to convince me that there's far more to the universe than the mere human eye can see. There are gaps.

When we look at a table, for instance, we think we're looking at something with mass. We are, of course, but there's actually more space in a table than there is mass, more space than there are particles. When we look at a distant star, we think, based on the evidence from our eyes, that we're seeing it now, but in fact it could have burned out years, even millennia, earlier. All that's left may be that last ray of light traveling through space from the era when it actually shone, and that could have been thousands, even millions of years ago.

Seeing a star right now? The speed of light is fast, but not that fast. What happens when we do that exercise, not with a grain elevator half a mile away, but with a star in a distant galaxy?

Whew!

That anomaly in time, of thinking we are seeing something now with the human eye, when in reality we are seeing something in the distant past, is only the first of so many gaps in time. Eventually time disappears altogether. Time, like space, becomes nothing. Once Einstein showed that time was not absolute, as Newton's three-dimension theory postulated, but relative, because of the four-dimension space-time view, it was not long until he was able to conclude that the past, the present, and the future all existed simultaneously. "The separation between past, present, and future is only an illusion, although a convincing one," he wrote.

His theory of relativity hinted that the faster we travel across space, the slower time progresses. If we could travel at the speed of light, time would cease altogether and we would exist only in timelessness. No time or eternal time might then be the same thing, just as in space, no thing and infinity are the same. That space traveler would have "popped out" of time and out of the illusion, the veil, altogether.

Remarkably, the arrow of time that we all perceive as moving exclusively forward toward the future is not supported by any of the known laws of physics. Time, and the consequent change in matter toward or away from entropy, can run in either direction, forward or backward, without violating physical law. It is only our rational perception, our need to perpetuate the illusion, perhaps, that insists time can only run forward.

Einstein knew this. Richard Feynman knew this. Stephen Hawking knows this. Plato probably knew this potential coincidence of no time and eternity as well. "Time is a moving image of eternity," he wrote. Most of the rest of the physics crowd is, like the rest of us, stuck in the illusion, stuck in the universe, stuck in the body.

The human body actually has a pretty limited view of things, it turns out. Our eyeballs can only perceive a fraction of the entire electromagnetic spectrum. The narrow band of light frequencies perceivable by the human eye stretch from 390 to 780 nanometers. Light carries the image onto our retina, a two-dimensional surface, and that two-dimensional image is then processed into a three-dimensional rendering, just as a three-dimensional hologram is projected from a two-dimensional holographic strip. Although a full quarter of the human brain is needed to process our visual input, humans still see far less than many inhabitants of the animal world.

We can see a rainbow, but we can't see radio waves

or X-rays or gamma rays. We'll never know it when Buzz Lightyear offs us. We'll never see it coming.

And yet we can see a dream. How is that?

Our vantage point within the veil, from within our bodies, is pretty limited. No wonder so many discoveries about our universe, even when proven later by physical evidence, started off as some physicist's dream—even before it appeared as chalk markings on a blackboard.

The physical *vis-à-vis* the mental: it's the gap that makes Escher's infinite waterfall or Magritte's vase of clouds so titillating. Of course there are other dimensions besides the fundamental four, even if so far they have only been "discovered" on that superstring physicist's blackboard.

Relying on the body's perceptions, then, to discover what we are doing here—who and what we are—has proven pretty much to be a lost cause. Worse yet, a body is noticeably finite within the fundamental four dimensions: it's mortal. Not only is it difficult to see what's going on, you've only got a couple years, in the grand scheme of things, to look.

And the things we can see are also finite. If something is perceivable, not just to our bodies, but even to our high-tech instrumentation like radio telescopes and microscopes, it must be some sort of particle or wave, something with some mass. Whether expanding or contracting, mass indicates a boundary. Ordered or disordered, particles always have perimeters.

No matter what direction we look in, no matter how sophisticated the "eyes" we look with, we'll always travel through the physical and mental universe to arrive at the ultimate gap, the final cosmic rabbit hole.

In today's physics this is known as a gravitational singularity, that point and moment either at the center of a black hole or at that first instant of the Big Bang, when no known tool for measurement, actual or theoretical, physical or mental, works.

It's the ultimate contrast. It's the gap of all physical gaps.

There may be space and matter in the physical universe, even dark matter and dark energy, but all of those things are somethings—not to sound too much like Dr. Seuss. And, yes, Diana had me read lots of Dr. Seuss and Lewis Carroll to "loosen me up." Make me a little lighter, a little more playful.

Beyond all those somethings, there must then be a nothing. Literally, a no thing.

In bodily terms, it's immortality. In physical universe

terms, it's infinity, but infinity is—in those same physical universe terms of something—the same as no thing. Both immortality and infinity are outside "thing-ness" altogether.

And it just may be that the ultimate no thing—that infinite wellspring of human consciousness—is me. And you.

When I came up behind Diana with my palms on her arms, we would for a moment produce our own little gap, together, and unite as one, as nothing, as everything. Love.

I never should have taken pride in hearing the question "Who is John Galt?" I should instead have asked the proper question, the one that Diana asked me: "*What* is John Galt?"

CHAPTER FIVE

∞◯◯∞

The dreamer can know no truth,
not even about his dream, except by awaking out of it.

—George Santayana

Nothing to something.

The path of an idea. The creation of a thought. The DNA of a dream. Each of these somethings comes out of a nothing.

Go ahead. Think a thought. It can be anything, from the cosmic to the mundane. " 'Good-bye' is 'God be with you'; 'Namaste' is 'I admire the Godhead within you' " or "I need to pick up more toilet paper at the supermarket tonight." Whatever it was, it was a creation, an idea that had never existed before. It came from that wellspring of your own consciousness, from you as a no thing.

Here's another nothing-to-something moment.

Find a blank space on a wall in the room you're in. Now imagine a simple red rose on the wall. Sure, maybe only *you* can see it, but you *can* see it. Same as your seeing your own dream. That rose is a creation. There was nothing there a second ago, and now there is. You can create it, place it on and take it off, at will. Now it's there. Now it isn't. Nothing to something. Then nothing again.

Are you imagining that rose? Of course you are. That's what the imagination is: the creation of something from nothing.

"Imagination is more important than knowledge," Albert Einstein noted. "I never came upon any of my discoveries through the process of rational thinking."

It's one thing to merely move a bunch of somethings around, as modern physics has been trying to do. It's altogether another thing to put something in a place where there was nothing before. The Parthenon, the Eiffel Tower, and the Burj Khalifa all started out in the imagination of the architect's mind, just like that red rose you just placed on your wall.

So, too, the *Mona Lisa*, Beethoven's Ninth Symphony, *War and Peace*, and the charmingly scrawled birthday card you received once from your four-year-old daughter.

So it is with a dream as well. Although parts of a dream may be drawn up from your memory, each dream you create is totally new. Nothing exactly like it has ever been created before. Each of your dreams is a something from a nothing. And look back at the most vivid dream you've ever had. How "real" was that?

Pretty real while it was happening, eh?

A creation from you, who are both nothing and infinite in your creative potential. You could create new thoughts, new ideas, new dreams all day long if you wanted to.

I didn't recognize it, but I had created a dream about myself. The dream that I was something. Diana relieved me sweetly, gently, and irrevocably from that illusion.

How is that thought, that idea, that red rose on the wall, that dream measured? By time, yes, perhaps, but are the dimensions we usually use to describe our physical universe—height, length, and width—adequate to precisely measure these creations?

How big or small is an idea? How much space does it take up? How much space can you make it take up?

Perhaps the vibrating strings of the fifth through the eleventh dimensions will prove to be more valuable in the measurement of these initially wispy creations.

Perhaps an idea, a dream, a vision passes up through these extra dimensions, one by one, from the nothingness of human consciousness as it manifests itself into a more and more substantial, perceivable idea, a more solid dream. From the idea to the chalk on the blackboard to the George Washington Bridge.

Perhaps those dimensions can be used to measure the "size" of our individual dreams.

Perhaps they also can measure the size of our collective dream: the veil.

"All human beings are dream beings. Dreaming ties all mankind together," wrote Jack Kerouac. That's right, Jack. It certainly does.

In fact, what if the entire perceived world is a dream? A collective dream. The veil itself. Maya. The ultimate illusion. The created playing field we put there so we could have a game.

How would we know?

Diana fed me more than just a diet of quantum physics and gravitational singularities. She stopped me from being so smart. She made me curious instead. She got me to dump my pride about my mind and let—ah, like a breath of fresh air!—uncertainty in.

"I do not know whether I was then a man dreaming I

was a butterfly, or whether I am now a butterfly, dreaming I am a man," said the Chinese philosopher Zhuangzi.

How would we know?

From the gaps.

All those little red flags we've been exploring here and that Diana waved before my eyes until *my own* eyes, not my body's eyes, finally opened. All those anomalies to the "rules of the game" that contradict our intellect, boggle our minds, and stand out as experiences that suggest a portal to Alice's unexpectedly cosmic rabbit hole.

Like extradimensional vibrations of superstring activity, these gaps confound the normal flow of life around us. Usually the gaps are purely personal—a little déjà vu, a little touch of ESP when a recently deceased relative assures us they're okay—but when they occur broadly, collectively, they become religion. They become miracles. It's the only way for our intellect to explain them, and the explanation stops the puzzling, stops the urge to find out what actually happened, stops the desire to peek behind the curtain or under the veil so we can actually see what's going on. Stops the uncertainty.

Explaining stops us from being afraid of getting an answer we might not want to face. What if we weren't "some thing"? What if we were only "no thing"? Too scary.

St. Francis's ability to communicate with animals is a good example of a miracle, for example, and frankly we're better off, we're saner, by continuing to wonder how the heck he did that. Our innate curiosity about life is too often blunted by explanations, especially religious ones, which, especially in the West, hide those pesky episodes where the gap in the veil is revealed.

It was not just Yoda or the Linda Blair of movies who levitated, for instance. Jesus walked on water. Buddha did too. In fact, so did St. Francis, St. Teresa of Avila, and Ignatius of Loyola, per contemporaneous reports.

Miracles?

Scientific minds usually attribute such events to trickery and illusion, of course. Nobody likes a gap. Certainly not scientists.

Religion "explains" gaps. Science dismisses them.

So what do we do when contemporary yogi masters or African shamans have been filmed levitating? What do we do with Daniel Dunglas Home, who was observed levitating out the third-story window of one building and back in through another? A host of witnesses reported him performing more than fifty levitations with his own body and levitating more objects around him, including chairs and tables. I wonder if he had that same mischievous grin on his face while levitating that the Cheshire Cat had.

And what about ghosts? What about the supernatural? What about resurrection?

Even scientists—bless their logical little hearts, and I used to be one of them—claim that man may only be using 7 percent of his mind at any given moment. What are we doing with the other 93 percent then? What is the potential there? If that 7 percent includes our logic, then are there not super-logical powers above mere reason, powers better measured by vibrating extradimensional strings than by the rational conclusions of length, width, depth, and time?

In fact, might these gaps we observe in the world around us—miracles, déjà vu, ghosts—not be gaps in the world around us, but really only gaps in our understanding of it? Inadequacies in our own logical understanding of the world around us?

Einstein's ranking of imagination over knowledge parallels the Taoist's mistrust of logic as well. We could also make a case that Jesus and Buddha's parallel teachings about loving your fellow man point to a far higher power than the West's obsession with the reward of heaven and the East's desire to eliminate the need for an endless cycle of deaths and rebirths. Both versions of immortality lie outside the veil. Outside the dream we may have created so we could have a game.

Heaven and Nirvana are both nongame states.

And what's the first rule of a successful game? Pretending it isn't a game. Otherwise one would always know the outcome. The first necessity of having a dream? Pretending it isn't a dream.

Are we pretending this isn't a veil?

If so, we do that pretty well…except for these pesky little gaps.

CHAPTER SIX

Some truths are just too big for people to face.

—George Orwell

The first gap that Diana ever noticed, she told me, was when she was a little girl attending Presbyterian church services with her parents one Sunday morning. She loved the church—the white wood slats on the outside, which always seemed freshly painted, and the polished oak pews inside, which always seemed to have been freshly varnished. No gaps there.

What she didn't understand, though, was that if you were going to go to heaven (or hell) after your body died and remain there as a spirit forever—in other words, if you were immortal—then how come you weren't immortal *before* you lived?

Our body's mortality, she reasoned, implied an

afterlife that is immortal. In the East, of course, the afterlife is also conceived of as a before-life. A person exists perpetually, as a soul, participating in one reincarnation after another. In the West, however, and in Christianity, the afterlife is said only to be an afterlife: the eternal residence of a soul in heaven or, worst case scenario, hell.

Was it possible, though, to have one-way immortality? Diana wondered. From here on out only?

Even at nine years old, that got her to thinking. Like Alice, her first step down that cosmic rabbit hole was to explore.

Oddly enough, until recently that one-way immortality was mirrored in some versions of the Big Bang theory. It posited one moment in time when the universe was born, and when nothing existed before it.

Perhaps creationists and evolutionists have more in common, cosmically, than they might have thought.

Today, however, after the discovery of gravitational singularities at the centers of black holes, science itself has abandoned any notion of one-way immortality. A gravitational singularity is a location space and in time at the center of a black hole or at the earliest moment of the Big Bang where man's conventional measurements of space and time cease to be relevant or to be contained within any coordinate system.

In short, there appears to be an infinity, or a no thing—same thing, really, in physical universe terms—at the start of the Big Bang or at the center of a black hole. The time-space gravitational pull is so intense there that neither general relativity terms nor quantum physics terms can be used to describe what occurs. In a black hole, time-space collapses in on itself; just before the Big Bang, it expands out exponentially.

That singularity is where infinity and no thing coincide.

In religious terms, of course, that would be immortality. That would be eternal life.

In dream terms, that would be the dreamer.

The most recent theoretical discoveries of contemporary physics also have more to say about the dream, the veil, maya, the grand illusion.

"The universe is a hologram," says theoretical physicist Leonard Susskind. He and other physicists in his camp have "proved" it with a tidy mathematical certainty.

Their research is derived from the study of black holes, from those extradimensional anomalies in the time-space continuum. Susskind says that as you pass into a black hole, you reach a "point of no return," a point at which the centripetal gravitational pull is so great that nothing can escape.

You pass that point of no return into a different "version" of space-time because the point of no return itself acts as a two-dimensional, holographic strip that projects a three-dimensional hologram into the interior of the black hole. It's the same principle as the retina of the human eye holding a two-dimensional image that is then transformed into a three-dimensional view of the world that we "see." At that point of no return in the black hole, though, you pass through what seems to be a scrambled hologram of everything that's inside the black hole—everything that is unrecognizable to those outside the point of no return. From the outside, none of it makes sense.

Alice, too, disappears into a world of distorted, impossible, charming creatures...but they seem so only to us. To Alice they're utterly real and their behavior is not questioned.

Contemporary physicists can show how everything in the universe can be seen as information—ideas or concepts transformed to tiny, encoded bits of data—and how that information flows off that two-dimensional holographic strip into the black hole. That point of no return, as you pass over the top of the gravitational waterfall and begin to cascade into the center of a black hole, contains the basic information about what is inside the black hole itself, just as a two-dimensional holographic strip contains the data necessary for the construction of a three-dimensional hologram.

Susskind has also discovered that the maximum amount of information in a region of space is proportional—not to the volume of the space that contains it, interestingly enough—to the perimeter area of that space. Therefore, the point of no return surrounding the universe, right at the edge of infinity or no thing, is also a two-dimensional hologram strip of informational bits that project the three-dimensional interior of the universe we live in.

It's the DNA of the illusion, of maya, of the veil, of the dream. Of the universe as hologram.

"In the beginning was the word, and the word was with God and the word was God" (John 1:1). Or, in twentieth-century terms, "In the beginning was the informational bit..." The holographic DNA for the creation of our universe.

The illusion we perceive. The dream we are dreaming. The maya we create.

"In the beginning was the word…" The word is a concept, the finest wavelength possible, and one word is just this side of nothing. The word might have been a minute vibration, a one-dimensional string, the eleventh dimension. Right up against immortality, infinity, no thing.

The word was the first act of the imagination.

That point of no return is also the point where, although you created the rules of the game and you're about to start the clock, you must forget that it is a game so that you can actually play it with intensity.

There is truth, and there is reality. Reality is an illusion because it is a truth that exists within time, even though the "greater" truth exists outside time, outside space, in infinity and as no thing.

Maybe we have those moments of déjà vu because those experiences really did happen before, when we first dreamed them, before they became part of the collective dream, part of our current reality.

Even the twentieth-century's greatest physicist, Einstein, knew this. Remember that he wrote, "Reality is merely an illusion, albeit a very persistent one."

This life is a mere sliver of a greater reality. Only the gaps in the veil point the way to the route—whether you think of it in spiritual or mathematical terms—back up out of the rabbit hole we've created.

My whole relationship with Diana, my whole time with her, was one big déjà vu. With Dagny, too—although for a while I had thought I loved her and had thought she would be an ideal mate for me, an equal mind—there was a feeling of déjà vu, but it was a "this-universe" déjà vu. We must have had something in common previously that had bound us together, probably in industry or science.

But Diana…Diana and I went way back before the veil, before the dream, before the illusion. We had cosmic déjà vu.

I had never thought about illusions or immortality—one way or the other—until I came back to Iowa to see Diana. I only knew from that first moment of grand déjà vu when she served me that first cup of coffee in that Iowa diner that she was a woman I was not going to be able to forget. A woman that I somehow had always remembered.

We must have been partners in the original crime, the original act of our imagination, before this universe.

It ended up placing us together in a little white cottage on the edge of a prairie in Iowa. True soul mates.

When Diana and I embraced, my hands on her arms, my chin on her head, we moved quietly and smoothly out of the illusion, out of the veil, and we returned "home."

Love.

I had studied Einstein, of course, at Patrick Henry University, but I had looked at him as someone who had gained new knowledge about the physical universe, about the mechanics of putting things together. His discoveries into the way things worked helped me to figure out how to build my static electricity engine.

It was Diana, though, who introduced me to a different Einstein, to an Einstein who, right alongside Alice, was scurrying down a rabbit hole to find out what was at the other end. He had found a gap in man's knowledge. Despite all his research into the operation of the physical universe, despite all his inquiry into the mechanics of the field we are playing in, Einstein knew that knowledge, per se, is only worth so much in the grand scheme of things.

"Knowledge is limited to all we now know and understand," he said, "while imagination embraces the entire world, and all there ever will be to know and understand."

Besides deriving the formula for energy, $E=MC^2$, Einstein had also derived the "formula" for creating an illusion. As George Bernard Shaw said, "Imagination is the beginning of creation. You imagine what you desire, you will what you imagine and at last you create what you will." And what we have imagined, willed, and created, apparently, is that peskily persistent illusion we have decided to recognize as reality.

It is so ironic. We put the illusion of a playing field there so we'd have something to play on, but we had to then pretend we didn't know it was an illusion just so we could play our games on it. One of our most important games, the physical science game, became to know what the illusion was composed of in the first place.

It's an unwinnable game, in a sense. The best kind.

Yet we all have such a deep, vibrant yearning to go back outside the game, to go "home."

CHAPTER SEVEN

This free will business is a bit terrifying anyway.
It's almost pleasanter to obey, and make the most of it.

—Ugo Betti

Diana once asked me, "If we have forfeited our ability to 'know' that this is all a game, have we also forfeited our free will in order to play it?"

"Is this another one of your Zen koans?" I asked. She said it wasn't.

If there were no free will, she asked me, wouldn't all of our actions be meaningless? There would be no true morality. On the other hand, if there were free will, how would we know what actions to take?

I was off into my next gap, but I didn't realize the second, implied aspect to Diana's question. She was beginning to expose the polar nature of the mind itself.

Instead, I rushed off into my rational research about free will.

The idea of free will has always run up against the idea of a deterministic universe; that is, a universe where everything has already been predetermined, either by God or by the natural order of things. A deterministic universe would, of course, invalidate the idea of us having free will or, at best, make the notion of our free will itself an illusion, like some divine placebo.

Both religion and science have contributed to our acceptance of a deterministic universe, one where there could be no free will. It's as if they figured out the inside track to this veil we're living in. If God is omnipotent and omniscient, some religious philosophers argue, then even though God has given man free will to act, the very fact that the outcome of any man's actions is known by God beforehand provides a deterministic underpinning to life that renders man's free will at best an illusion, at worst a cruel trick.

In Islam and in many Christian beliefs, the greater free will of God always trumps the illusory free will of man.

In the West, only the Jews have proposed that the part of a man's spirit that does innately have free will is that part which is joined to God. It is therefore independent of the physical playing field. It acts in unison with God's own free will.

In the East, many Hindus have also prescribed a deterministic view for each person's life by noting that the acts of one life determine one's status in the next life. Buddhist theories of reincarnation focus more on the moral determinism of one life to the next by stressing that the morality of one's actions in one life determine one's status in the next. Hence, any life lived has been predetermined by an earlier life. You have been on a course predetermined by each previous life...for how long now?

Until the advent of quantum physics, the Newtonian universe was also seen by science to be deterministic. The basic formulae derived from observation of physical phenomena accurately predicted or determined future events. Every time an apple broke from its stem on a tree, it fell to the ground. Hence, if we live in a universe where all the laws of activity are already set in motion and predetermined, how could there be free will? The pattern had already been set.

With the discovery of quantum physics, though, and the observation of thoroughly random motions of particles on a microscopic level, we came to understand that a completely indeterminate aspect governed all life. So indeterminate, in fact, that it, too, precluded any meaningful nature to free will. Ironically, after so many years of Newtonian determinism, the philosophical conclusions drawn about free will now by science were not those of a newly potent free will in a now indeterminate universe—they were just the opposite.

If life is full of random, unpredictable motion, then there can be no free will. You can never confidently predict the consequences of your actions. Think of it this way: when your ice cream scoop topples out of the cone you are holding, it can as easily end up on your eyebrows as on your chin.

The more I puzzled over this and over Diana's question to me, the more I seemed to arrive at the fact that, if there was anything even resembling free will, it was only an illusion. Despite the fact that *we* may have set up this illusion, this game of games, we may have inadvertently condemned ourselves to a near-eternity of predetermined experience.

This is a pretty nihilistic view of things. Life had suddenly become very serious for me again. Could I ever know whether my apparent free will was an illusion or an actuality? This was a problem I seemed to have no way to solve. I let Diana know I had reached a dead end.

"The solution to the problem is never on the level of the problem itself," she told me, then turned and walked back into the house, leaving me on the porch looking out at the rolling green landscape of the most fertile farmland in America. She had once heard that quote from Maharishi Mahesh Yogi, she told me. I believe she also thought that the Beatles could never be wrong.

This problem, like all rational problems, I realized, was based on a contradiction, a contrast, a polarity. Either that's the way the universe actually was, or the tool we were using to explore the universe, the mind, was not equipped to analyze it. That conflict itself—mind or universe—was the first facet of the whole multifaceted problem of free will.

What, then, was the "other level" I needed to resolve the problem?

The gap in between, I realized. Then, as the landscape around me brightened, Diana stuck her head around the porch screen door and asked me, "Are you aware of looking at those green fields?"

"Yes."

"And are you aware of the thoughts you're holding about free will?"

"Yes."

"Then doesn't that awareness precede what you're aware of—thoughts or hills?"

"Yes again, Diana."

"And what is it that's being aware of those green fields, those thoughts?" she asked with a little smirk as she quietly closed the screen door and went back to making one of the most perfect pie crusts I would ever taste.

"Oh! Me."

I remembered a quote by Tolstoy that I had read in the previous couple days, a quote I found searching for the answer to her original question about free will, a quote I had quickly dismissed for being too childish, too simple-minded, too solipsistic to be valuable: "You say: I am not free. But I have raised and lowered my arm. Everyone understands that this illogical answer is an irrefutable proof of freedom."

The solution to the problem is never on the level of the problem itself. The proof of free will, like the proof of many of the most important things in life, is ultimately subjective, not objective. It is known by the self, not by the thoughts or the universe the mind is speculating about.

Could it really be that simple?

Guess it was, because I "popped out"—big time—on that one. I was the porch, the grass contour of the rolling green fields, the rich loam of the fertile land. I was the full blue sky and the white, white clouds. I was more, even, than I had been in back of my father's garage, when I was the tower and my body. I was fully me.

I raised and lowered my arm. I had free will.

What was the alternative conclusion? That my awareness itself was an illusion? That my awareness itself was invalid? Those negative thoughts were only the mind's ideas about awareness. Awareness itself was an experience. My experience.

Sorry, Descartes. Ain't no "I think, therefore I am." I am, and sometimes I choose to think. And sometimes I just choose to raise and lower my arm.

I realized, as well, that part of the problem here is language, one of the mind's main tools. Language is composed of words. Words are symbols. They stand for things or ideas about things. What if the things they stand for don't exist—other than in the mind of the thinker?

Take Freud's "ego" or "id." Can we look around and find one of these things, or does the word itself only stand for an idea in Sigmund's cocaine-addled mind? What about a singularity? Dark energy? Dark matter? Are they actually tangibly out there, or are they mere ideas on the blackboard of some physicist's mind? Mere symbols for a referent that does not exist?

What if we have become so enamored of the mind—just like I and my band of destroyers had—that we've forgotten that subjective experience is paramount in establishing truth? We've let the mind itself create a most enticing illusion: the veil of rational thought. The god of logic. The touchstone of what we should or should not believe. We've trapped ourselves in the polar operations of the mind itself. The only saving grace is that that polarity offers up, ultimately and by necessity, a gap between the poles.

Science itself may have determined that the universe is a hologram, but, ironically, science itself may only be a metaphor.

So, the good news is that we do have free will. The bad news, as they also say, is that we do have free will. Ugo Betti had it pegged. If we have free will, we have to be responsible. The horror.

We have to face the tornados that rip across the Iowa plains. The hurricanes and typhoons that decimate coastal populations year in and year out. Lightning strikes. Flash floods. A volcano's searing lava. Lethal epidemics. And those are just the *natural* disasters.

What about war, or even the threat of war from third-world countries on the cusp of having nuclear arms? What about school shootings? Terrorist attacks? Pederasts? Drug addicts? Even petty criminals, from whom no person's wallet or household possessions are safe? What about the IRS?

If we live in a dream, we may also have to live through nightmares. If we have a game, we have antinomies. We have contradictions. We have opposites. If we think we have good guys, we have to have bad guys. That's the way it works. Either in real life or in our mind's interpretation of real life.

Is it possible to look at any horrific event—a man butchering a cocker spaniel pup, let's say—and see it merely as a sequence of actions, of impersonal activities without meaning, without emotion, without value? Without projecting something of ourselves on it?

Perhaps. Perhaps not. Mentally, good guys and bad guys are taken to be an integral part of the game.

The experience of compassion is taken as an integral, emotional part of the game. Sure, it has its opposite—callousness. A logical, mental look leads to the conclusion that both exist, but feeling compassion is also an experience beyond any rational thought, and itself may be one other indicator of our having free will. Would we ever spontaneously feel compassion if it were not possible to help another truly?

I raise my arm. I lower it.

I soothe a baby's brow with my palm. I hug a grieving widow. I pull a maniac with a knife off a helpless cocker spaniel. I choose to be a good guy.

I love.

When I didn't know, couldn't see and was oblivious to the gaps, I was a kind of bad guy. I was a destroyer. The one who wanted to stop the motor of the world. The one who enlisted the best minds in the country to do the same. A bad guy. Not as bad a guy as Hitler. Not as bad as Charlie Manson. But kind of a bad guy.

Now, with help from Diana, I can look, I can see. I can be a good guy again.

CHAPTER EIGHT

Illusion is the first of all pleasures.

—Voltaire

For me, the ultimate persuasion that this is all an illusion happens when I experience my separateness from it. When I "pop out." It's not just an out-of-body experience, but an out-of-universe one.

I first became convinced I was looking at an illusion as I looked at the world around me from Diana's porch—at the rolling green fields of Iowa, the vast blue sky and clean white clouds above. I thought I just might want to give up on life. After all, if it was just an illusion, why bother to do anything?

But, in fact, a funny thing happens when you stare at everything around you as if it were an illusion. Try it. Yes, at first the landscape seems eerily fixed and unmoving—a still hologram, as it were—but after just a few minutes, rather than feel hopeless or apathetic about it all, you feel utterly and simply bored. You don't have a game.

What is boredom other than an emotional damper hobbling our deep and native urge to act? We want a game. We want something to do. Staring at the "still" hologram for a minute or two actually prompts you to act, to do something, to have activity of some kind or another, mental or physical.

This deep, innate urge may be in harmony with our original purpose—as immortal, omniscient and cosmically bored beings—to have a game. To put up the playing field. To create the illusion.

Add to that the certainty that you have free will—not just as an intellectual construct, but as an experience—and you're off and running.

There is, though, another layer that is far more difficult to recognize as contributing to the illusion. The mind. And that was the second, implied part of Diana's question to me about free will.

The mind always wants to know, but the mind—with its primal, dual aspects of good and evil, black and white, positive and negative just like the binary informational coding of Susskind's holographic strip surrounding the universe—can never "know" the illusion. While the East has taken the route of the Zen koan to explore the endless workings of the mind and give us the experience of our actual selves, popped out, the West has embedded itself in the workings of the mind, through science and math, to figure everything out.

Of course, that is a great game in itself: using our mental faculties, refining them more and more precisely, to figure the universe out. It might actually be the best of games because it's one that can never be won. It's endless. We'll never be bored playing it. It may be the grandest of illusions, all in itself.

A fellow I ran into one afternoon at the gas station in our little Iowa town—I always remember his name was Zach because I'd never met anyone named Zach before—said something very interesting to me. He was something of an inventor himself, and we'd gotten to talking about physics and astrophysics and space exploration when he suddenly stopped me in mid-sentence.

"Space," he said, "although perpetually expanding and mysterious, is not and never has been the 'final frontier.' The human mind remains the singular frontier in need of exploration."

I guess I was primed for this thought by my studies with Diana, and I realized he was right.

He finished filling his tank, said good-bye, and drove off out of town like some mythic Clint Eastwood character who'd just saved the town from the bad guys. A déjà vu moment for me, for sure.

I started thinking about that and about the tools of the mind.

I realized Susskind's conclusion that the universe is a hologram is proven. It is true, by tidy mathematical certainty, as are other twentieth- and twentieth-first-century breakthroughs in quantum physics and astrophysics. But mathematics is a language like any other. It possesses all the assets, as well as the liabilities, of any other language.

Take the sentence "Ten million blue elephants flew across the sky in a Wedgwood teacup." Unlike "id" or "superego," symbols that may only refer to an idea in Sigmund Freud's mind, each word in this sentence has a specific referent in the real world. The grammar and syntax of the sequence of words is also consistent with the rules of the English language itself. The sentence should therefore make sense, but it doesn't.

Only our experience tells us that this sentence rings

THE DÉJÀ VU EXPERIMENT

untrue. We cannot figure this out from the sentence itself, or from the rules by which the sentence was composed. We only know it is untrue because we know, by experience that there are no elephants that fly, that there are no elephants that are blue, and that there are no elephants teeny enough to allow ten million of them to fit in a Wedgwood teacup. Similarly, we only know we have free will from the experience of raising our arm.

Mathematical sentences—formulae and calculations—are similar: they can "make sense" as symbols put together correctly according to their language's own internal rules of operation, but they may not accurately describe any phenomena in the physical universe itself. Even if we remove the possibility that some of those symbols themselves might only stand for things that exist in an astrophysicist's mind—maybe "dark matter" and "dark energy" are like Freud's "ego." Even when each mathematical symbol stands for a function known to exist, and when the syntax of the equation is perfect, they still may not advance our knowledge of the world around us.

Take fractals, for example. These are structures or geometric shapes that have a symmetry of scale. No matter how large or how small the iteration of a fractal's basic pattern or design, each rendition looks the same as all the others. Fractals, as formal mathematical equations, came about after mathematician Benoit Mandelbrot began wondering how to measure the geometry of the irregular shapes we find all around us in nature: the coast of Britain, for example, or the surface area of a cauliflower. They *should* be able to be measured, but odd things begin to happen when we try.

Fractals had been observed in nature forever, of course, in the decreasing spirals of a conch shell or in the repeated symmetrical patterns found in pineapples, snowflakes, or lightning branches. These design repetitions have often been used by artists and can be admired in paintings such as Salvador Dalí's *Visage of War* or Augusto Giacometti's alpine flowers series, as well as in the ornate repetitions of line in Aubrey Beardsley's pen-and-ink drawings (and in the creations of all those who have mimicked him since). Even William Blake's engravings have a fractal reiteration of harmonious shapes in them; his fascination with the symmetrical in nature and the creator of it all comes through in the two lines from his poem "The Tyger": "What immortal hand or eye/Could shape thy fearful symmetry?"

In short, fractals, as part of the illusion itself, have been giving us pleasure as art for centuries.

Mandelbrot's calculations to measure irregular geometries progressed from science, from what could be seen—as did Giacometti's alpine flowers—to what could be imagined. He noticed that the more closely you measure, the greater the length or area being measured becomes, until mathematically the length becomes infinite. That's similar to Zeno's paradox about halving the measured distance to a wall each time you take a step toward it. You will never arrive at the wall itself because the distance to the wall expands, mathematically, to infinity.

Mandelbrot's mathematical syntax also verges into another representation that is beyond anything that can be conceived of by the five senses: dimensions appeared "between" the conventional three dimensions of height, length, and breadth as you calculate the lines and areas of fractal compositions with mathematical equations. Even with equations as tidy as Susskind's that describe the universe as a hologram.

These odd, in-between fractal dimensions begin to parallel the mathematical configurations of string theory's dimensions five through eleven and the holographic identity of Susskind's universe. Yet they are, like the ten million blue elephants, beyond anything "provable" by human experience, although they are sound when defined by mathematical lingo. They may help define the gap that scientists are investigating to help them understand the physical universe. It's their version— bless them—of following Alice down the rabbit hole, but even though they seem real, they may not be actual truth.

Could all of these calculations solely be the result of inquisitive minds bent on playing a game they can never win?

When we judge these theories by our own experience of the physical universe, the syntactically perfect equations seem relegated to the land of flying blue elephants. But, on the other hand, they have yielded plenty of practical, tangible results: heat exchangers, digital imaging, photographic enlargement, and video game design are just some of the fields that have all been enhanced by fractal mathematics. Just as my own extrapolation of Van de Graaff's calculations on static electricity generators culminated in my static electricity engine.

Man does benefit. These scientists are good guys.

Could it also be that these anomalies, these gaps between the mind-generated and the physical-universe-generated, are also portals to the multiverses that are the "logical" conclusions to the mathematical formulae of fractals, Big Bang singularities, and superstring theory dimensions five through eleven? So many of the current mathematical models of the universe point in the direction of multiverses, William James's term for the hypothetical but mathematically sound set of other universes that may exist parallel to our own.

Have we set up, in our infinite caution never to be bored, never to not have a game, even more illusions than the veil we're currently involved in?

Whether the universe is described as an infinite stretch of finite patterns that repeat, fractal-like, into a necessary occurrence of other universes parallel to ours; or as a continuing sequence of Big Bang singularities bubbling up into alternate, parallel universes; or as strata of brane universes floating in and out of the additional dimensions—all the syntactically and grammatically correct mathematical "sentences" that define it still suggest the greatest likelihood is that other universes exist.

Following the text of contemporary physics and

mathematics down their respective rabbit holes in their attempt to "fill in the gaps" of an incompletely understood universe and its rules of operation, all roads seem to lead to infinity. But infinity, just like no thing, only exists outside of everything. It existed before we set up this game. Or any other game.

Some time ago, Diana introduced my originally scientific mind to Jorge Luis Borges's story "The Garden of Forking Paths," which describes an infinite labyrinth of time and choice, and to Escher's infinite staircases, which portray the possibility of an infinite geometry. Contemporary mathematical texts do the same.

For all three of these, though, the litmus test for truth can only be our own experience.

And our most profound experience is our experience of the gap. Without explanation. Without embellishment. Without interpretation. Without justification. Pure.

The déjà vu experiment, the great game that we are playing, which we have made almost impossible to recognize as a game, is revealed to us only by those moments when we experience the gaps that link us back to that before-game reality.

There is the "little" déjà vu, for instance, when we

feel we've experienced some actual thing before.

Then there's the "big" déjà vu, when we are linked timelessly to our pregame existence of infinity or no thing. In physical universe terms, it was when all possibilities for creation still existed, be they in one or more universes.

When we "pop out."

When we are the dreamers without a dream.

CHAPTER NINE

Blind but sure-footed,
we step forward as if into a remembered dance.

—Margaret Atwood

I am nearing death.

I have to say, I do wonder what the "gap" in death might be. Maybe there is no gap. Maybe as the body slips away into death, the veil fades and the dream collapses. Just as when you are starting to awake from sleep there is a moment when you're aware of your dream, and of the fact that you are dreaming.

Or maybe one just "pops out," ready for the next adventure. Maybe death is the gap, itself.

I've never really feared death. For a long time, when I was acting the destroyer, I had feared not having enough time left to do what I wanted to do. But understanding as I do now that time is itself an illusion, that is no longer a concern for me.

Will I miss what's around me? The flowing green fields of Iowa with their magnificent summer thunderstorms? The children playing in the neighbors' yards? Sweet homemade peach ice cream on an August evening?

Maybe I'll miss these things. It probably depends on where I resurface and what I have in front of me then. Will I begin my next adventure in some other, parallel universe constructed of altogether different geometries and put together by an entirely different set of laws for its physics? Which of the many potential multiverses might I spring forth into, and what beauty might lie in their landscapes?

Or maybe I'll end up right back here.

More important, what will I want to accomplish? What purposes will I put forth for my new identity in a new illusion? What goals will I have for my next scenes in this cosmic play? How will I align them with our original goals for this grand illusion? Our stage is set for endless characters, endless landscapes, endless games full of life, death, good, evil, and ugliness and beauty. We never can know whether any of these are part of a game if we are to play it "for real."

Except for the gaps.

Buddha was right. Suffering, too, is an illusion. We are eternal spirits. In this world of both joy and pain, this controlled experiment, we can do no permanent harm to each other. Our current concept of time is an illusion. But we can cause pain to others with our actions and words, and therefore we do have a duty to frame our experiences with our fellow man with morally acceptable actions.

My body is wearing out. There's nothing horribly wrong with it, but I can tell it's nearing the end of the line. I didn't expect it to last forever. No one else's has. But this one has served me well.

It occurred to me when Diana passed on, years ago, that all human anguish is based on having a body. Wars, crimes, even accidents result in bodily death. Tornados, tsunamis, and floods destroy bodies individually or in multitudes. Even the death of others—loved ones and family—cause emotional pain by removing a body from our day-to-day lives, especially if it was one we had grown fond of.

I wonder if we made bodies part of the original equation in this illusion, or if they came after. Because of them, there appears to be pain, suffering, and anguish.

Or maybe pain, suffering, and anguish are the necessary antinomies to joy, elation, and ecstasy in a binary universe. Not because we dreamed up bodies, but because we dreamed up minds. Bodies have certainly become the most dramatic way, though, of participating on the stage of this almost endless play.

I do still wonder how the mind fits into death, though. Will I leave it behind, like my body? Or will I bring it along with me like some piece of quantum baggage for my next adventure? It may be partially a burden to carry, with its number of bad memories, and partially a container of the necessities I'll still want with me: logic, rationality, language, inquisitiveness.

My guess is the latter.

Soon I will know.

I'll go to sleep one night and the next morning I will no longer wake up as John Galt. I'll be somebody else. Or something else. I'll be off on a new adventure.

Cool.

When I met Diana, my attention was stuck like a magnet to my immediate past. I worried about how wrong had I been, trying to stop the engine of the world. How much damage had I done to so many people by denying them, in some cases, the necessities of life, transportation, communication, and even a sense of hope for the future. I had made the world seem worse to them than it really was because I focused too much of their attention on what was wrong, not on what could be made right.

In short, I was serious.

I don't know that I became what you could call flippant or glib after my years under Diana's life tutelage, but I certainly became able to laugh at the immense seriousness with which I had taken life. With her guiding me to look at all of the many gaps in life that we notice in art, in science, in religion, and which point to a larger reality, I began slowly to move my attention from the past to the future.

As eternal spirits, we may be part of an endless controlled experiment of our own device—a sort of cosmic Hotel California—and there might not be any real, permanent harm in what we do. Today I can start planning my more long-range future, beyond this life, and start determining how I can best act to provide my fellow man with the most beneficial experience.

I can begin to figure out how to align my own long-range purposes in life—whatever and wherever I might find them, no matter in which "universe" I might find myself—with our own innate and original purpose in setting up this grand stage of infinite performances.

My attention, then, has been almost exclusively focused on my intentions and their potential for manifestation in the future, no matter what, materially, that future will be. My intentions, Diana has allowed me to see, are closest to my true eternal self and are most native to the true "home" we all share, the "home" that can be found, now and forever, by following the gaps we find down this peculiar rabbit hole we now inhabit.

If, in my passing, I could bequeath something, I would bequeath the choice to follow the gaps down that rabbit hole so we can exercise the choice, the free will, to measure our own acts by both our immediate and our eternal purposes as individuals acting in harmony with humanity.

We don't want Atlas holding up the world all by himself or shrugging and walking away from his mythic task. We want all of us to support the globe and share the task of shaping our future in a Nash equilibrium where the best choices are those that are best for humanity and for ourselves.

If time itself is an illusion, our deepest moments of choice span our entire existence, from before there was a game until after the game ends—if it ever does. Our most important choices, as we continue to create this crucible of illusion and force our individual and collective truth into it, can give each of us a voice that speaks to the benefit of all, mortally and eternally.

In my earlier concerns, though, when I was first with Diana, I did draw two significant conclusions about my past.

First, I was wrong about stopping the motor of the world, not just because stopping that motor may have been harmful to too many people, but because that was the wrong motor.

The motor that I should have put my efforts into stopping was the motor of the mind. *That's* the motor that runs the world. *That's* the motor that still, today, is running the world—poorly. Ignoring the gaps, going full speed ahead on a railway of fixed ideas, creating problems with incorrect solutions—the motor that thinks that way needs to be stopped. The collective mind travels at breakneck speed down the interstate, completely on automatic pilot.

That's the motor that is creating so much of the needless suffering, struggle, and anguish in the world today. War in the Middle East, nuclear weapons in North

Korea, an epidemic of starvation in Ethiopia, genocide in Darfur and Syria. And all the while the young people who could change all this are playing video games incessantly in the plush suburbs of East Cincinnati instead of truly growing up.

The mind is merely a tool, like a screwdriver. It is not God any more than God is a screwdriver.

I don't want my legacy to be that of a destroyer, although I don't mind being seen a spear-shaker in the drama of this stage we call life. I'd rather be remembered for helping to usher in an era when ideology can be openly and creatively discussed. Discussed outside of the fixed confines of that mental interstate where thought occurs only on automatic pilot and with fixed ideas. I used to be amused when I heard those legends of my having found Atlantis, but now, at the end of this life's journey, I'd be happy to be thought of as a guide to the true Atlantis.

My second conclusion was that I had been right about one thing. It must be the individual who is given the most value. We can see by the shape the world is in that the collective conclusions of mankind are not working. It is only the individual who will ever be able to recognize a gap and then to use his mind to fasten the screws that are now so loosely out of place.

Unfortunately, the electromagnetic waves of every big-screen TV and small-screen iPad are keeping hypnotized the minds of those who should be seeing life from a fresh angle. They must come to understand that we don't know everything—that, in fact, we hardly know anything. Someone must teach them that the sometimes fearful uncertainty of not knowing what life is about is far less scary than the blind certainty of "knowing" what life is about when you're wrong. We need to exit permanently those thirty miles of robotic interstate we're traveling, no matter how frightening the new route might seem and how uncharted our maps might be of the surrounding territory of our universe.

"I think it's much more interesting to live not knowing than to have answers which might be wrong," my hero Richard Feynman once said. He did not feel frightened by not knowing things.

What, then, have I learned in my journey down the rabbit hole following Alice—and my love, Diana—on the trail of gaps that lead to the fulfillment of that wise Greek admonition to "know thyself"?

I've learned that the first step, the step that will enable you to spot the first gap when it happens, is to recognize that you don't usually know yourself in the truest sense. If you're lost in the woods, the very first step is not to roar off in any old direction as fast as you can, it's to recognize instead that you're lost. Only by accepting that certainty can you then have a hope of finding your way out.

So, too, with "knowing thyself." First step is to recognize you probably don't.

Sure, it's scary. It's disorienting. It's difficult to do. You've got to look for yourself and trust what you see. By doing that, you're going to discover a gap.

Guaranteed. And that will be your first certainty.

The certainty of the gap.

CPSIA information can be obtained
at www.ICGtesting.com
Printed in the USA
LVOW04*1319111215
466353LV00006B/15/P

9 780989 718608